FOREWORD
by Captain R B Middleton FNI President The Nautical Institute.

The confidential marine accident reports which we publish each month in our journal SEAWAYS have indicated that mariners are frequently confronted with hazardous situations relating to the COLREGS.

To investigate the problems The Nautical Institute initiated in 2002 an international survey amongst seagoing officers to enquire if they thought there was a problem, what the causes could be and to suggest solutions.

The results were alarming with nearly 50% of the respondents stating that in their view the problems were caused by ignorance and wilful disregard of the rules.

However respondents also drew attention to other factors like distractions due to GMDSS false alarms, VHF chatter, VTS reporting, and paperwork.

When it came to solutions respondents highlighted both the need to improve the education and training of watchkeeping officers and better supervision by senior officers on board.

Undoubtedly we all have to work to improve standards of bridge watchkeeping and this thoughtful guide produced by the North of England P&I Club is an excellent place to start.

Officers can work through a page at a time discussing the elements with colleagues and relating the contents to the full form of the rules.

For experienced officers there is an opportunity to work through the text as a refresher and to use the booklet as a training aid when coaching junior officers and cadets on watch.

It is our duty as nautical professionals to ensure that we are responsible at all times for the safe navigation of our ships and this booklet provides a valuable text to re-establish good practice.

INTRODUCTION

CONVENTION ON THE INTERNATIONAL REGULATIONS FOR PREVENTING COLLISIONS AT SEA, 1972 (COLREGS)

This guide is based on the COLREGS poster series produced by the North of England P&I Association's Risk Management Department between 2001 and 2004.

The aim of these posters is not to provide an all encompassing guide to the Rules but to dispel some frequent misunderstandings and misinterpretations of the most widely used Rules and to provide a graphic illustration of the correct steps to decipher and to apply the Regulations.

While it is vital to know the contents of the Regulations, simply being able to recite them verbatim is no substitute for knowing how to apply them. A coherent full working knowledge of the text and its application is vital.

The UK Marine Accident Investigation Branch has stressed the importance of Rule 2 and their concern that this vital Rule is so often overlooked.

Rule 2 - Responsibility

(a) *Nothing in these Rules shall exonerate any vessel, or the owner, master or crew thereof, from the consequences of any neglect to comply with these Rules or of neglect of any precaution which may be required by the ordinary practice of seaman, or by the special circumstances of the case.*

(b) *In construing and complying with these Rules due regard shall be had to all dangers of navigation and collision and to any special circumstances, including the limitations of the vessels involved, which may make a departure from these Rules necessary to avoid immediate danger.*

This fundamental feature of the COLREGS highlights the fact that you, the mariners, are responsible for your own actions and you have to comply with the Rules while ensuring that you take all precautions of the ordinary practice of seafarers - there is no substitute for the application of common sense on the seas.

During the Nautical Institute's "Improving the application of the COLREGS" survey, it was noted by the late Captain François Baillod FNI, that reported incidents related to "uncertainty", "violations" and "disregard for the COLREGS". While many cited "ignorance", "poor knowledge of the COLREGS", or "lack of training" as reasons for the poor and ineffective application of the Regulations.

The Collision Regulations were devised to make the navigation of ships safer - however we must remember they are also the law and all should observe them. There can be no excuses - ignorance is no defence and if you disobey these laws there will be consequences. This guide is intended to provide a reminder of the Rules and the processes that should be applied in making them work.

Much of the guidance text has been provided by Captain Roger Syms who has enthusiastically embraced the idea of providing an easy to use guide to the COLREGS in conjunction with the original North of England poster series. This project has also been able to make use of the experiences expressed through the Nautical Institute's survey into the COLREGS.

The North of England also acknowledges with thanks, the contribution and suggestions of the UK Marine Accident Investigation Branch on the conceptual ideas behind the project.

Look Out

Rule 5 Look-out

Every vessel shall at all times maintain a proper look-out by sight and hearing as well as by all available means appropriate in the prevailing circumstances and conditions so as to make a full appraisal of the situation and of the risk of collision.

Rule 5 - Look Out!

"Look out" means pay attention to everything! - Not just looking ahead out of the window but all round the vessel, using all your senses and all appropriate equipment available to you.

KEEP AN EYE AND EAR ON EVERYTHING THAT GIVES YOU INFORMATION INCLUDING:

- **Radar/ARPA** - Be aware of the effects of clutter, of small targets and the range of the set.
- **Radio transmissions** - Keep an ear on what is going on in the vessel's vicinity.
- **Sound Signals** - Can you hear any sound signals near-by? Be aware of the effect of keeping a closed wheelhouse, and of distracting noises in a closed space.
- **Course and Position Navigation Aids** - If you have to alter course make sure there is safe water available.
- **Depth indicator** - frequently and systematically monitor the depth of water beneath the vessel.
- **Be aware** - VTS and AIS are there to help you.

BEWARE OF DISTRACTIONS!

- **Alarms** - Do not allow alarms, such as false GMDSS transmissions, to distract you from keeping a proper look-out.
- **Lights** - Do not allow lights on your vessel to impair your vision
- **Communications** - VHF transmissions and mobile phones are not your first priority. Do not allow them to distract you.

Rule 5 applies at all times - there must always be someone looking-out. If weather or conditions cause concern then more lookouts may be needed and should be called without hesitation.

Safe Speed

Rule 6 - Safe Speed

Every vessel shall at all times proceed at a safe speed so that she can take proper and effective action to avoid collision and be stopped within a distance appropriate to the prevailing circumstances and conditions.

In determining a safe speed the following factors shall be among those taken into account:

(a) By all vessels:
  (i) the state of visibility;
  (ii) the traffic density including concentrations of fishing vessels or any other vessels;
  (iii) the manoeuvrability of the vessel with special reference to stopping distance and turning ability in the prevailing conditions;
  (iv) at night the presence of background light such as from shore lights or from back scatter of her own lights;
  (v) the state of wind, sea and current, and the proximity of navigational hazards;
  (vi) the draught in relation to the available depth of water.

(b) Additionally, by vessels with operational radar:
  (i) the characteristics, efficiency and limitations of the radar equipment;
  (ii) any constraints imposed by the radar range scale in use;
  (iii) the effect on radar detection of the sea state, weather and other sources of interference;
  (iv) the possibility that small vessels, ice and other floating objects may not be detected by radar at an adequate range;
  (v) the number, location and movement of vessels detected by radar;
  (vi) the more exact assessment of the visibility that may be possible when radar is used to determine the range of vessels or other objects in the vicinity.

Rule 6 - Watch your Speed!

Rule 6 allows you to make your own judgement as to the most appropriate speed at any time for your vessel, taking into account the prevailing circumstances and conditions.

CAN YOU REACT EFFECTIVELY?

Without exception the safest speed is a reduced speed, because:

- **It allows you to stop or turn effectively**
- **It gives you more time to think and to act in a developing situation -**
  The time to react and respond effectively is all important - vessels moving too quickly can often fatally impair their own watchkeeper's risk assessment processes.
- **If collision does occur the resulting damage is likely to be a lot less**

Remember the radar or ARPA is not infallible. It may miss some targets altogether or it may show very large targets as weak echoes.

Navigational aids such as GPS can be equally suspect - don't rely on one instrument for information, double check it against others.

Constantly monitor your speed - The situation at sea is constantly changing and what can be deemed a safe speed in one situation can change with circumstances, sometimes suddenly!

Maintaining high speeds for commercial considerations should not be tolerated - it is no excuse or defence for proceeding at an unsafe speed.

Risk of Collision

Rule 7 - Risk of Collision

(a) Every vessel shall use all available means appropriate to the prevailing circumstances and conditions to determine if risk of collision exists. If there is any doubt such risk shall be deemed to exist.

(b) Proper use shall be made of radar equipment if fitted and operational, including long-range scanning to obtain early warning of risk of collision and radar plotting or equivalent systematic observation of detected objects.

(c) Assumptions shall not be made on the basis of scanty information, especially scanty radar information.

(d) In determining if risk of collision exists the following considerations shall be among those taken into account:

(i) Such risk shall be deemed to exist if the compass bearing of an approaching vessel does not appreciably change;

(ii) Such risk may sometimes exist even when an appreciable bearing change is evident, particularly when approaching a very large vessel or a tow or when approaching a vessel at close range.

Rule 7 - Watch that Ship!

Listen as well as look! - As with keeping a look out, you must use all the information and equipment available to you to determine risk of collision.

Use the compass to check the bearings of approaching vessels - compare it with the radar bearing.

If you have operational radar you must use it.

Are you using true or relative vectors? **If you have an Automatic Radar Plotting Aid (ARPA) you should use the RELATIVE VECTORS for determining risk of collision and TRUE VECTORS to ascertain the other vessel's actual movement.**

You must be aware of the limitations and use of the ARPA and interpret information displayed correctly.

If you are not fitted with an ARPA you must run a RADAR PLOT.

Is the target passing ahead or astern? Or is it going to collide? - Remember the primary information you need to answer these questions is relative information.

Don't trust ARPA to give you an accurate Closest Point of Approach (CPA). - Where possible take at least half a mile off each indication to be safe. (If it shows a CPA of half a mile assume it is collision).

Don't rely on change of bearing as an indicator of clearance. - As a target approaches it's change of bearing should speed up significantly. If the change in bearing does not accelerate, treat it as a danger.

Don't relax your vigilance - Keep monitoring the situation until the target is passed and well clear.

Action to Avoid Collision

Rule 8 – Action to Avoid Collision

(a) Any action to avoid collision shall be taken in accordance with the Rules of this part and shall, if the circumstances of the case admit, be positive, made in ample time and with due regard to the observance of good seamanship.

(b) Any alteration of course and/or speed to avoid collision shall, if the circumstances of the case admit, be large enough to be readily apparent to another vessel observing visually or by radar; a succession of small alterations of course and/or speed should be avoided.

(c) If there is sufficient sea-room, alteration of course alone may be the most effective action to avoid a close-quarters situation provided that it is made in good time, is substantial and does not result in another close-quarters situation.

(d) Action taken to avoid collision with another vessel shall be such as to result in passing at a safe distance. The effectiveness of the action shall be carefully checked until the other vessel is finally past and clear.

(e) If necessary to avoid collision or allow more time to assess the situation, a vessel shall slacken her speed or take all way off by stopping or reversing her means of propulsion.

(f) (i) A vessel which, by any of these Rules, is required not to impede the passage or safe passage of another vessel shall, when required by the circumstances of the case, take early action to allow sufficient sea-room for the safe passage of the other vessel.

(ii) A vessel required not to impede the passage or safe passage of another vessel is not relieved of this obligation if approaching the other vessel so as to involve risk of collision and shall, when taking action, have full regard to the action which may be required by the Rules of this Part.

(iii) A vessel the passage of which is not to be impeded remains fully obliged to comply with the Rules of this Part when the two vessels are approaching one another so as to involve risk of collision.

Rule 8 - Do Something and do it Early!

Remember a positive alteration made very early on is better than a large panic alteration at the last minute! - the closer you are to the other vessel the more you will have to do to avoid collision.

Don't judge a "positive" and "ample" action just by the amount of alteration - confirm it by the change in CPA. Keep checking the situation until the risk of collision is past and clear.

Make your actions obvious to other vessels - Small alterations of course are dangerous; they don't usually solve the problem and don't give the other vessel a clear indication of what you are doing.

Use the engines - If your ability to alter is constrained then **SLOW DOWN** or **STOP.**

The other vessel may also be obliged to take action. Always bear in mind what that action may be.

NOT IMPEDING?

- If the Rules require you "not to impede", it means you must make a very early alteration to make sure risk of collision does not develop.
- If you are the "not to be impeded" vessel be prepared for the other vessel not to take the correct action. If a collision is imminent you have to act however constrained you are!

Think about what you are doing - Actions taken to avoid collision should follow the observance of good seamanship - this is where we have to apply professional and sound practical judgement!

Use the Trial Manoeuvre setting on your ARPA, if available.

Traffic Separation Schemes

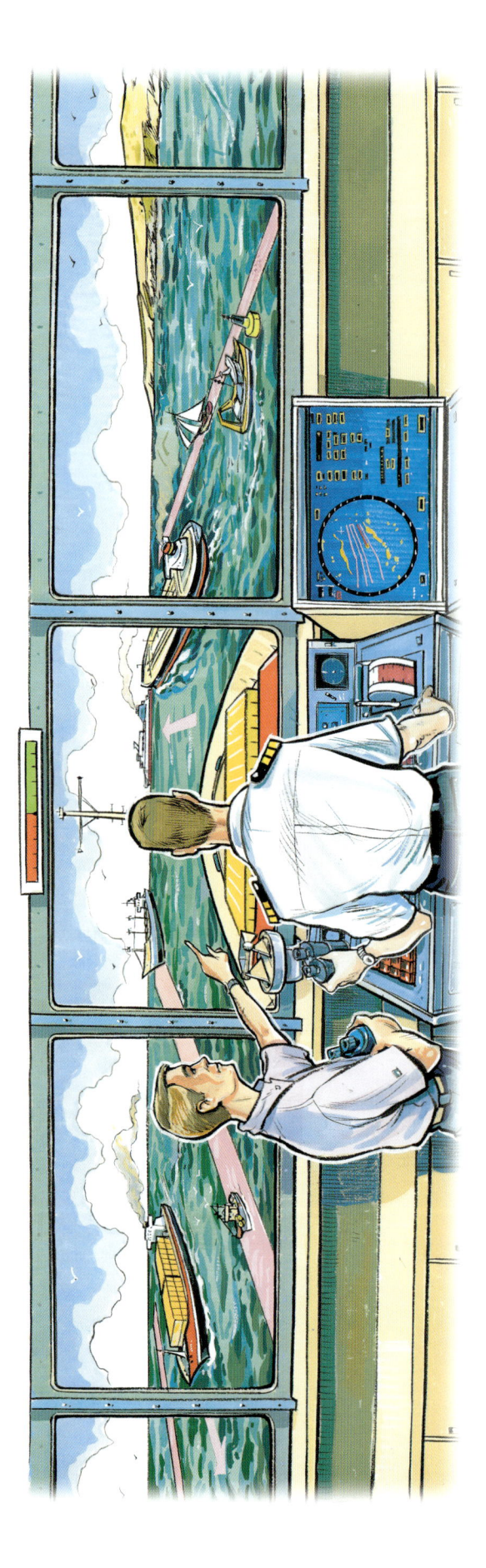

Rule 10 - Traffic Separation Schemes

(a) This rule applies to traffic separation schemes adopted by the Organization and does not relieve any vessel of her obligation under any other rule.

(b) A vessel using a traffic separation scheme shall:
 (i) proceed in the appropriate traffic lane in the general direction of traffic flow for that lane;
 (ii) so far as practicable keep clear of a traffic separation line or separation zone;
 (iii) normally join or leave a traffic lane at the termination of the lane, but when joining or leaving from either side shall do so at as small an angle to the general direction of traffic flow as practicable.

(c) A vessel shall, so far as practicable, avoid crossing traffic lanes but if obliged to do so shall cross on a heading as nearly as practicable at right angles to the general direction of traffic flow.

(d) (i) A vessel shall not use an inshore traffic zone when she can safely use the appropriate traffic lane within the adjacent traffic separation scheme. However, vessels of less than 20m in length, sailing vessels and vessels engaged in fishing may use the inshore traffic zone.
 (ii) Notwithstanding subparagraph (d)(i), a vessel may use an inshore traffic zone when en route to or from a port, offshore installation or structure, pilot station or any other place situated within the inshore traffic zone, or to avoid immediate danger.

(e) A vessel other than a crossing vessel or a vessel joining or leaving a lane shall not normally enter a separation zone or cross a separation line except:
 (i) in cases of emergency to avoid immediate danger;
 (ii) to engage in fishing within a separation zone.

(f) A vessel navigating in areas near the terminations of traffic separation schemes shall do so with particular caution.

(g) A vessel shall so far as practicable avoid anchoring in a traffic separation scheme or in areas near its terminations.

(h) A vessel not using a traffic separation scheme shall avoid it by as wide a margin as is practicable.

(i) A vessel engaged in fishing shall not impede the passage of any vessel following a traffic lane.

(j) A vessel of less than 20m in length or a sailing vessel shall not impede the safe passage of a power-driven vessel following a traffic lane.

(k) A vessel restricted in her ability to manoeuvre when engaged in an operation for the maintenance of safety of navigation in a traffic separation scheme is exempted from complying with this Rule to the extent necessary to carry out the operation.

(l) A vessel restricted in her ability to manoeuvre when engaged in an operation for the laying, servicing or picking up of a submarine cable, within a traffic separation scheme, is exempted from complying with this Rule to the extent necessary to carry out the operation.

Rule 10 - Traffic Separation Lanes are not Rights of Way!

There is no right of way - Just because you are navigating within a Traffic Separation Scheme (TSS) does not give you right of way over other vessels.

The other COLREGS continue to apply within a TSS.

Remain within the lanes - but if your alteration for a crossing vessel is likely to take you outside the scheme this does not exempt you from following the Rules. If you are not happy about it SLOW DOWN or STOP.

BE WARY!
Actions of vessels navigating in the vicinity of a TSS can be UNPREDICTABLE.

Look out for crossing vessels on the edge of the scheme. They may alter to cross at right angles or they may alter parallel to the scheme to find a less crowded place to cross.

Look out for High Speed Craft (HSC) - HSC tend to present collision risks wide on the beam. HSC do often alter course early, however you cannot assume they will always do so.

Overtaking

Rule 13 - Overtaking

(a) Notwithstanding anything contained in the Rules of Part B, Sections I and II, any vessel overtaking any other shall keep out of the way of the vessel being overtaken.

(b) A vessel shall be deemed to be overtaking when coming up with a another vessel from a direction more than 22.5 degrees abaft her beam, that is, in such a position with reference to the vessel she is overtaking, that at night she would be able to see only the sternlight of that vessel but neither of her sidelights.

(c) When a vessel is in any doubt as to whether she is overtaking another, she shall assume that this is the case and act accordingly.

(d) Any subsequent alteration of the bearing between the two vessels shall not make the overtaking vessel a crossing vessel within the meaning of these Rules or relieve her of the duty of keeping clear of the overtaken vessel until she is finally past and clear.

Rule 13 - Leave Ample Room!

Don't forget! If you are not sure you are an overtaking vessel, you must assume that you are and keep clear.

Don't pass close - overtaking invariably takes time, so make sure you have a safe distance between you and the other vessel. *(Where possible this should be at least greater than your hard-over turning circle.)*

Beware of interaction! - if you are forced by traffic to pass closer be very careful that interaction does not occur.

Avoid crossing ahead - If you are not on parallel courses and passing clear, cross astern rather than ahead.

Does the ship being overtaken know you are there? Always assume they do not!

Remain vigilant - Remember, you remain an overtaking vessel until you are finally passed and clear.

Head-on Situation

Rule 14 - Head-on Situation

(a) When two power-driven vessels are meeting on reciprocal or nearly reciprocal courses so as to involve risk of collision each shall alter her course to starboard so that each shall pass on the port side of the other.

(b) Such a situation shall be deemed to exist when a vessel sees the other ahead or nearly ahead and by night she could see the masthead lights of the other in line or nearly in line and/or both sidelights and by day she observes the corresponding aspect of the other vessel.

(c) When a vessel is in any doubt as to whether such a situation exists she shall assume that it does exist and act accordingly.

Rule 14 - Ship Ahead!

"Nearly reciprocal" does not mean exactly right ahead. If a vessel is ahead and coming the other way on an opposite course and roughly within half a point (6 or 7 degrees) of either side of the bow, Rule 14 applies.

IF YOU ARE STILL NOT SURE ASSUME A HEAD-ON SITUATION ANYWAY AND ACT ACCORDINGLY!

UNDERSTAND RULE 8

- Rule 8(a) says, "any action to avoid collision shall be taken in accordance with the Rules of this Part ..." *("this Part" being the Steering and Sailing Rules)*
- It means that if you have a vessel fine to starboard, even at some distance away, altering to port to increase the clearance may be construed as not being in accordance with the COLREGS. You should always go to starboard as directed by the Rule.

Alter early! - Do not wait for the other vessel to act, the closer you get before taking action the greater the steps you subsequently have to take to avoid collision.

Crossing Situation

Rule 15 - Crossing Situation

When two power-driven vessels are crossing so as to involve risk of collision, the vessel which has the other on her own starboard side shall keep out of the way and shall, if the circumstances of the case admit, avoid crossing ahead of the other vessel.

Rule 15 - Watch Vessels to Starboard!

Give way early - If you are the give-way vessel, take early action so the other vessel knows your intentions.

Avoid crossing ahead - Go to starboard, astern of the vessel if at all possible.

Be considerate - If the other vessel is hampered in any way, action must be taken even earlier to reassure the other vessel.

Be positive! - If in doubt over crossing or being overtaken, assume you are crossing and keep clear.

Use your engines - Remember that you should always have the option of using your engines as well as your helm.

Action by Give-way Vessel

Rule 16 - Action by Give-way Vessel

Every vessel which is directed to keep out of the way of another vessel shall, so far as possible, take early and substantial action to keep well clear.

Rule 16 - Keep Everybody Happy!

DO NOT HESITATE! - Alter early and adequately enough to show the other vessel clearly what you are doing.

Confirm your alteration - make sure that your action has had the desired effect by checking the increase in the CPA on the radar.

KEEP EVERYBODY HAPPY!

- Following Rule 16 makes life easier for all vessels.
- If the watchkeeper on the other vessel is happy with the situation and with your alteration there is less likelihood of he/she doing anything unexpected.
- Treat other vessels as you would like to be treated yourself. Apply some courtesy to seamanship and common sense.

Action by Stand-on Vessel

Rule 17 - Action by Stand-on Vessel

(a) (i) Where one of two vessels is to keep out of the way the other shall keep her course and speed.

   (ii) The latter vessel may, however, take action to avoid collision by her manoeuvre alone, as soon as it becomes apparent to her that the vessel required to keep out of the way is not taking appropriate action in compliance with these Rules.

(b) When, from any cause, the vessel required to keep her course and speed finds herself so close that collision cannot be avoided by the action of the give-way vessel alone, she shall take such action as will best aid to avoid collision.

(c) A power-driven vessel which takes action in a crossing situation in accordance with subparagraph (a)(ii) of this Rule to avoid collision with another power-driven vessel shall, if the circumstances of the case admit, not alter course to port for a vessel on her own port side.

(d) This Rule does not relieve the give-way vessel of her obligation to keep out of the way.

Rule 17 - Is She Altering?

THERE ARE TWO STAGES TO RULE 17:

- **17(a)(ii) At some distance off** - when "as soon as it becomes apparent ... that the vessel required to keep out of the way is not taking appropriate action...", you may take your own action to avoid collision.

    PROVIDED you do not alter to port for a vessel on your port side in a crossing situation.

- **17(b) At close quarters** - when "collision cannot be avoided by the give-way vessel alone", you should take the best action you can to avoid collision.

Conduct of Vessels in Restricted Visibility

Rule 19 – Conduct of Vessels in Restricted Visibility

(a) This rule applies to vessels not in sight of one another when navigating in or near an area of restricted visibility.

(b) Every vessel shall proceed at a safe speed adapted to the prevailing circumstances and conditions of restricted visibility. A power-driven vessel shall have her engines ready for immediate manoeuvre.

(c) Every vessel shall have due regard to the prevailing circumstances and conditions of restricted visibility when complying with the Rules of Section I of this Part.

(d) A vessel which detects by radar alone the presence of another vessel shall determine if a close-quarters situation is developing and/or risk of collision exists. If so, she shall take avoiding action in ample time, provided that when such action consists of an alteration of course, so far as possible the following shall be avoided:

   (i) An alteration of course to port for a vessel forward of the beam, other than for a vessel being overtaken;

   (ii) An alteration of course towards a vessel abeam or abaft the beam.

(e) Except where it has been determined that a risk of collision does not exist, every vessel which hears apparently forward of her beam the fog signal of another vessel, or which cannot avoid a close-quarters situation with another vessel forward of her beam, shall reduce her speed to the minimum at which she can be kept on her course. She shall if necessary take all her way off and in any event navigate with extreme caution until danger of collision is over.

Rule 19 - Restricted Visibility Changes Everything!

Rule 19 is a different set of rules that only apply when "vessels are not in sight of one another".

THERE IS NO SUCH THING AS A STAND-ON VESSEL IN RESTRICTED VISIBILITY. Every vessel must act!

Avoid altering to port (19(d)(i)) - An alteration of course to port (should be avoided) for a vessel forward of the beam, except for a vessel you are overtaking.

Avoid altering towards (19(d)(ii)) - If a target presents a collision or close quarters risk **abeam or abaft the beam** you must act but avoid altering towards it.

THIS DOES NOT MEAN you cannot alter towards ANY vessel that may be abeam or abaft the beam. Rule 19(d)(ii) only applies if there is a collision or close quarters risk.

Use your engines - Again always remember that you should have the option of using your engines as well as your helm.

Proceed at a safe speed - have your engines available for immediate manoeuvring.

If you don't have radar, for whatever reason, you must proceed with extreme caution. Listen for sound signals and respond accordingly.

REMEMBER!

The **COLREGS** are not just advice to the mariner - they are **THE LAW**.

If you disobey the law you will suffer the consequences.

Ignorance of the law is never a defence.

COLREGS
A GUIDE TO GOOD PRACTICE

The COLREGS are the foundations upon which the safe navigation and conduct of vessels are built.

It is vital that all who work with the Rules have a full and detailed knowledge of not only the text but of all the elements to effectively apply them.

The North of England P&I Association's Risk Management Department has in recent years produced a series of posters to highlight the Rules and this publication brings together the full set of posters combined with additional guidance on often misunderstood and neglected facets of their application.

Roger Syms' maritime career has spanned some 47 years, from deep-sea tankers to Hoverlloyd and Hoverspeed, as Captain and Flight Manager. He also has a degree in Nautical Studies from Plymouth Polytechnic.

He has assisted with the development of regional VTS systems and has lectured at the Australian Maritime College. While at the College he oversaw the design, installation and commissioning of new large-scale simulation facilities.

In 1998 he worked on the upgrade of Australian certificates to STCW'95. This provided a rare insight into the views of different nationalities to the COLREGS.

Since 2002 he has been heavily involved in the Nautical Institute's strategic COLREGS project.

The North of England P&I Association, with offices in the United Kingdom, Greece and Hong Kong, is a leading international mutual marine liability insurer with in excess of 45 million GT of entered tonnage.

Founded in 1860, the Association has long recognised the importance of providing loss prevention advice to its Members believing this to be the most effective way to reduce the number and scale of claims. The Association has developed a worldwide reputation for the quality and diversity of its loss prevention initiatives.

£30
ISBN 0-9542012-9-9

NORTH OF ENGLAND